TABLE OF CONTENTS

AT A GLANCE

TABLE OF CONTENTS	3
TEACHER ASSESSMENT RUBRIC	4
STUDENT SELF-ASSESSMENT RUBRIC	5
INTRODUCTION	6

ROCK TYPES
Teacher Notes — 7
Student Activities — 8

FOSSILS
Teacher Notes — 13
Student Activities — 14

EXPLORING MINERALS
Teacher Notes — 17
Student Activities — 18

MINING
Teacher Notes — 22
Student Activities — 23

FUN WITH ROCKS
Teacher Notes — 26
Student Activities — 27

EXPERIMENTING WITH EROSION
Teacher Notes — 30
Student Activities — 32

EROSION PREVENTION!
Teacher Notes — 42
Student Activities — 43

WEATHER PATTERNS
Teacher Notes — 47
Student Activities — 49

PRECIPITATION
Teacher Notes — 55
Student Activities — 56

WEATHER INSTRUMENTS
Teacher Notes — 62
Student Activities — 63

WASTE
Teacher Notes — 71
Student Activities — 73

MANAGING WASTE
Teacher Notes — 79
Student Activities — 80

REDUCE, REUSE, RECYCLE!
Teacher Notes — 87
Student Activities — 88

Teacher Assessment Rubric

Student's Name: _____   Date: _____

Success Criteria	Level 1	Level 2	Level 3	Level 4
Knowledge and Understanding Content				
Demonstrate an understanding of the concepts, ideas, terminology definitions, procedures and the safe use of equipment and materials	Demonstrates limited knowledge and understanding of the content	Demonstrates some knowledge and understanding of the content	Demonstrates considerable knowledge and understanding of the content	Demonstrates thorough knowledge and understanding of the content
Thinking Skills and Investigation Process				
Develop hypothesis, formulate questions, select strategies, plan an investigation	Uses planning and critical thinking skills with limited effectiveness	Uses planning and critical thinking skills with some effectiveness	Uses planning and critical thinking skills with considerable effectiveness	Uses planning and critical thinking skills with a high degree of effectiveness
Gather and record data, and make observations, using safety equipment	Uses investigative processing skills with limited effectiveness	Uses investigative processing skills with some effectiveness	Uses investigative processing skills with considerable effectiveness	Uses investigative processing skills with a high degree of effectiveness
Communication				
Organize and communicate ideas and information in oral, visual, and/or written forms	Organizes and communicates ideas and information with limited effectiveness	Organizes and communicates ideas and information with some effectiveness	Organizes and communicates ideas and information with considerable effectiveness	Organizes and communicates ideas and information with a high degree of effectiveness
Use science and technology vocabulary in the communication of ideas and information	Uses vocabulary and terminology with limited effectiveness	Uses vocabulary and terminology with some effectiveness	Uses vocabulary and terminology with considerable effectiveness	Uses vocabulary and terminology with a high degree of effectiveness
Application of Knowledge and Skills to Society and Environment				
Apply knowledge and skills to make connections between science and technology to society and the environment	Makes connections with limited effectiveness	Makes connections with some effectiveness	Makes connections with considerable effectiveness	Makes connections with a high degree of effectiveness
Propose action plans to address problems relating to science and technology, society, and environment	Proposes action plans with limited effectiveness	Proposes action plans with some effectiveness	Proposes action plans with considerable effectiveness	Proposes action plans with a high degree of effectiveness

Student Self Assessment Rubric

Name: _____ Date: _____

Put a check mark ✓ in the box that best describes you.

Expectations	Always	Almost Always	Sometimes	Needs Improvement
I am a good listener.				
I followed the directions.				
I stayed on task and finished on time.				
I remembered safety.				
My writing is neat.				
My pictures are neat and colored.				
I reported the results of my experiment.				
I discussed the results of my experiment.				
I know what I am good at.				
I know what I need to work on.				

1. I liked _____
   _____

2. I learned _____
   _____

3. I want to learn more about _____
   _____

Teacher Notes

INTRODUCTION

The activities in this book have two intentions: to teach concepts related to earth and space science and to provide students the opportunity to apply necessary skills needed for mastery of science and technology curriculum objectives.

Throughout the experiments, the scientific method is used. The scientific method is an investigative process which follows five steps to guide students to discover if evidence supports a hypothesis.

1. **Consider a question to investigate.**
   For each experiment, a question is provided for students to consider. For example, "Do rocks have magnetic properties?"

2. **Predict what you think will happen.**
   A hypothesis is an educated guess about the answer to the question being investigated. For example, "I believe that certain rocks contain minerals that are magnetic". A group discussion is ideal at this point.

3. **Create a plan or procedure to investigate the hypothesis.**
   The plan will include a list of materials and a list of steps to follow. It forms the "experiment".

4. **Record all the observations of the investigation.**
   Results may be recorded in written, table, or picture form.

5. **Draw a conclusion.**
   Do the results support the hypothesis? Encourage students to share their conclusions with their classmates, or in a large group discussion format.

The experiments in this book fall under thirteen topics that relate to three aspects of earth and space science: **Rocks, Minerals, and Erosion, Weather,** and **Waste and Our World.** In each section you will find teacher notes designed to provide you guidance with the learning intention, the success criteria, materials needed, a lesson outline, as well as provide some insight on what results to expect when the experiments are conducted. Suggestions for differentiation are also included so that all students can be successful in the learning environment.

Assessment and Evaluation:

Students can complete the Student Self-Assessment Rubric in order to determine their own strengths and areas for improvement. Assessment can be determined by observation of student participation in the investigation process. The classroom teacher can refer to the Teacher Assessment Rubric and complete it for each student to determine if the success criteria outlined in the lesson plan has been achieved. Determining an overall level of success for evaluation purposes can be done by viewing each student's rubric to see what level of achievement predominantly appears throughout the rubric.

ROCK TYPES

Teacher Notes

LEARNING INTENTION:
Students will learn about different rock types and discover the types that are where they live.

SUCCESS CRITERIA:
- describe the different types of rocks
- locate and illustrate four different rock types in the neighborhood
- identify rocks as either igneous, sedimentary, or metamorphic
- describe the characteristics of each of the rock types, relate these to their origins

MATERIALS NEEDED:
- a copy of "Types of Rocks" Worksheet 1, 2, and 3 for each student
- a copy of "What's Rocking Your Neighborhood?" Worksheet 4 and 5 for each student
- access to different areas in the neighborhood in order to collect different types of rocks
- plastic bags or small cardboard boxes (one per student)
- magnifying glasses (one per pair of students)
- clipboards, pencils (one for each student)
- chart paper, markers

PROCEDURE:
*This lesson can be done as one long lesson, or in two or three shorter lessons.

1. Using Worksheets 1, 2, and 3, do a shared reading activity with the students. This will allow for reading practice and learning how to break down word parts in order to read the larger words in the text. Along with the content, discussion of certain vocabulary words would be of benefit for students to fully understand the passage.

Some interesting vocabulary words are:
- igneous
- sedimentary
- metamorphic
- pressure
- magma
- weathering
- compact
- volcanic eruption

2. Explain to students that they will go on an exploration to discover some rock types that are in their neighborhood. Give each student a plastic bag or small cardboard box. Instruct them to collect about four interesting rocks that they would like to bring back to the classroom. **(These rocks will become a classroom rock collection, to be used throughout the unit.)**

3. Return to the classroom. Give students Worksheet 4. Options for this activity:
   - students can work individually using the rocks that they collected
   - students can work in a small group, pooling their rocks together, then choosing four rocks to illustrate on Worksheet 4

4. Students will visit the classroom rock collection to choose a rock from each type – igneous, sedimentary, and metamorphic. Using Worksheet 5, they will describe in detail the characteristics of the rocks and explain how these relate to the origin of the rocks.

DIFFERENTIATION:
Slower learners may benefit by participating in step 4 as outlined in the procedure section, but instead, choosing one rock each to bring to the small group that will work with teacher support to complete this activity on one shared piece of chart paper that will outline each of the rock types' characteristics and origins.

For enrichment, faster learners could write about which rock type they think is most common in their neighborhood, explaining their choice.

OTM2155   ISBN: 9781770789630
© On The Mark Press

Types of Rocks

There are three different types of rocks. These types are igneous rocks, sedimentary rocks, and metamorphic rocks. Let's read more about them!

Igneous

Igneous rocks have been formed by heat. This type of rock forms when molten rock, called magma, rises to the earth's surface through cracks caused by earthquakes and volcanic eruptions. When magma reaches the earth's surface, it becomes hot lava. This lava will cool into cold hard rock, which is known as molten rock. Igneous rocks are formed.

Basalt

Usually these rocks are heavy, but there are some that have air bubbles, which makes them very lightweight. ▼

Granite

The two most common igneous rocks are granite and basalt. Granite is often used for kitchen countertops. Basalt is sometimes used in construction of buildings, and in paving stones.

Sedimentary

Rocks are broken up by a natural process called "weathering". Weathering happens when rocks are washed away by water or blown away by wind, causing them to be broken into smaller pieces. These smaller pieces of rock settle into layers on land, in riverbeds, lakes, and oceans.

The weight of the layers of sand and water presses down on the bottom layer and squeezes out the water, which compacts it. The grains of sand and tiny rocks begin to stick together, eventually becoming stone. This is **sedimentary** rock.

Sandstone, limestone, and shale are the common types of sedimentary rock.

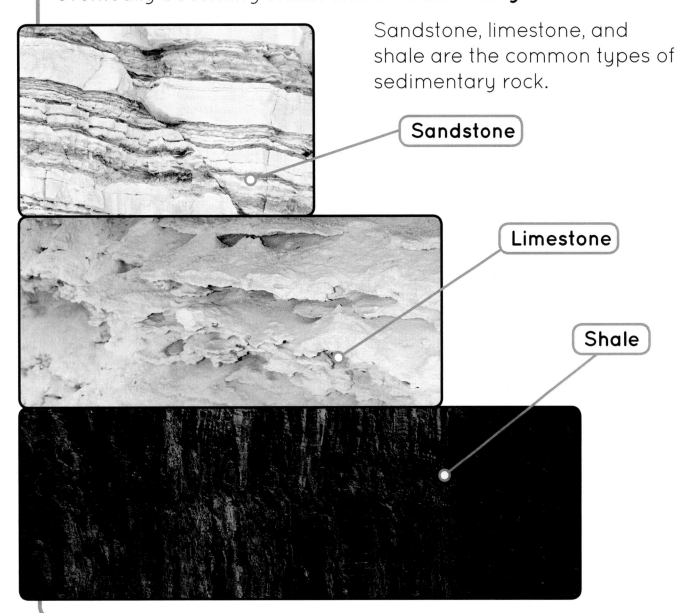

Metamorphic

Metamorphic means a change in form. Metamorphic rocks were all once igneous or sedimentary rocks. Hot masses of magma sometimes slowly push out of the earth's crust. The heat and pressure of the magma invades an igneous or sedimentary rock, causing it to change its form.

When limestone, a soft sedimentary rock is invaded by a mass of magma, it is changed into marble, a very hard stone.

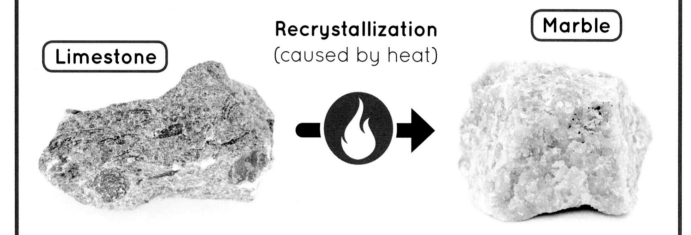

Did you know?

The Acasta Gneiss is a large metamorphic rock outcrop in the Northwest Territories, Canada. This rock outcrop is located about 300 kilometres north of Yellowknife. It is believed to be the oldest known rock in the world, about 4 billion years old!

Worksheet 4   Name:

What's Rocking Your Neighborhood?

You know a lot about the different types of rocks. Do you wonder what types of rocks are in your neighborhood? Let's go on a rock exploration!

In the boxes below, illustrate some rocks that you find. Label them as **igneous,** sedimentary**,** or **metamorphic**.

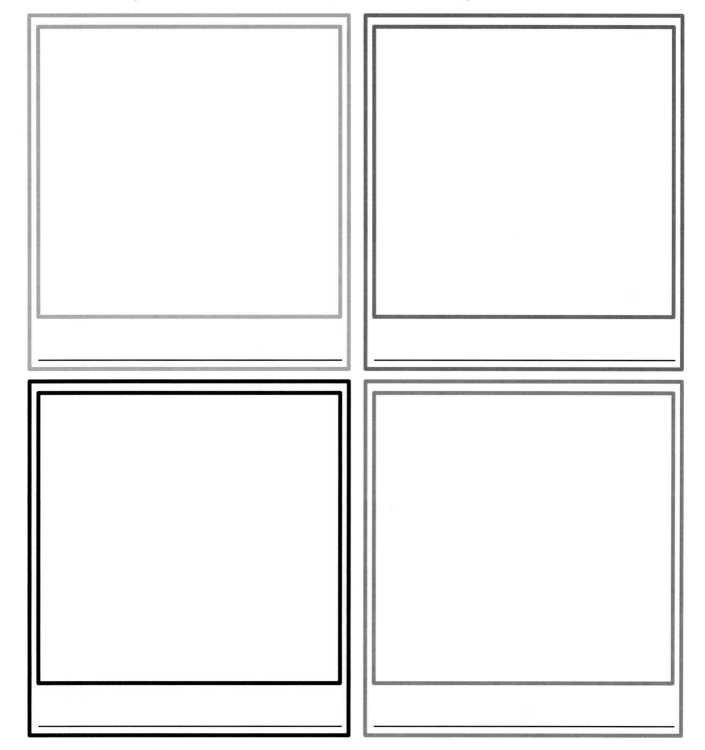

Worksheet 5	Name:

Visit your classroom rock collection. Choose a rock sample from each of the rock types.

A) **Describe its characteristics.**
- What does it look like?
- How does it feel?

B) **Explain its origins.**
- How do you think it was formed?

Rock Type Characteristics	Origins
Igneous • • • •	_____ _____ _____ _____ _____
Sedimentary • • • •	_____ _____ _____ _____ _____
Metamorphic • • • •	_____ _____ _____ _____ _____

FOSSILS

Teacher Notes

LEARNING INTENTION:
Students will learn about fossil formation and how they help us to understand Earth's history.

SUCCESS CRITERIA:
- identify a fossil and describe its features
- describe how a fossil is formed in sedimentary rock
- explain how fossils help us to understand the Earth's history
- create a model of a fossil

MATERIALS NEEDED:
- a copy of "Fossils" Worksheet 1 for each student
- a copy of "Going on a Fossil Hunt!" Worksheet 2 for each student
- a copy of "Making a Fossil" Worksheet 3 for each student
- access to different areas in the neighborhood in order to locate fossils or have samples of fossils in the classroom readily available for student exploration
- modeling clay, dish soap, newspaper, plaster of Paris, a jug, access to water
- a Popsicle stick, an aluminum pie plate, a disposable cup (a set for each student)
- some shells or small figurines like dinosaurs (these can be found at a dollar store)
- paint, paint brushes (optional)
- magnifying glasses, pencils (for each student)

PROCEDURE:
*This lesson can be done as one long lesson, or be divided into two shorter lessons.

1. Using Worksheet 1, do a shared reading activity with the students. This will allow for reading practice and learning how to break down word parts in order to read the larger words in the text. Along with the content, discussion of certain vocabulary words would be of benefit for students to fully understand the passage.

   Some interesting vocabulary words are:
   - fossil
   - remains
   - sediment
   - decay
   - paleontologist
   - mould
   - imprinted
   - compressed
   - seep
   - skeletons

2. Students will have an opportunity to locate a fossil. They can revisit the classroom rock collection, or go back out into the neighborhood to locate some, or you may provide them with some fossil samples to examine for this activity. Give them Worksheet 2 to complete. Afterwards, encourage students to share their fossil discovery with a peer in order to compare their findings. *For some extra fun, try an iPad app called "Earth Stories – Paleontology".

3. Explain to students that they will create a model of a fossil. Give them Worksheet 3 and the materials to begin their creation. Read through the materials needed and what to do sections to ensure their understanding of the task.

DIFFERENTIATION:
Slower learners may benefit by working with a peer to locate a fossil. Omit the written component of Worksheet 2, replace with a small group discussion, with teacher support, to explain where and how their fossil was formed.

For enrichment, faster learners could access the internet to research a paleontologist to learn about the discoveries he/she made and how this has helped us to understand Earth's history.

Fossils

What is a Fossil?

A fossil is like a mould of the remains of a plant or animal from a long time ago that has been imprinted in sedimentary rock. Fossils can be parts of bodies like bones, teeth, wings, or feathers. They can also be shells or plant leaves. Sometimes, they can be animal tracks.

How do Fossils Form?

When a plant or animal died a long time ago, part of it became trapped in sediment, like sand, mud, and small pieces of rock. Over time, it was compressed with more and more sediment. As decay happened, water with minerals seeped into the structure of the object, replacing it with rock like minerals. The object became fossilized and a part of sedimentary rock.

Did You Know?

A Paleontologist is a scientist who finds and uses fossils to learn about the plants and animals that were on Earth a long time ago. They have used fossil bones to make skeletons of dinosaurs and other animals. This helps us to know how large they were, how they walked, and what they ate.

Worksheet 2 Name:

Going on a Fossil Hunt!

Go on a fossil hunt! Examine the rocks that are in your classroom collection, or go back out into your neighborhood to look for evidence of a fossil formation!

Record your findings by illustrating and describing the fossil that you locate.

Fossil Specimen

_____
_____
_____

Where and **how** do think your fossil formed?

_____
_____
_____
_____

Making a Fossil!

You've read about them. You've hunted for them. Now it is time to create your very own fossil!

You'll need:

- modeling clay
- dish soap
- a Popsicle stick
- newspaper
- paint
- a disposable cup
- an aluminum pie plate
- a loose mix of plaster of Paris
- some shells, or small figurines (dinosaurs)
- paint brushes

What to do:

1. Line your table with newspaper.
2. Push a thin layer of modeling clay into the bottom of an aluminum pie plate, be sure that it covers the whole bottom.
3. Rub some dish soap on your figurine so that it doesn't stick to the clay.
4. Push your figurine into the clay so that it forms an imprint, then remove.
5. Using the Popsicle stick and disposable cup, mix up a loose mixture of water and plaster of Paris.
6. Slowly pour the plaster of Paris mixture into the aluminum pie plate, so that it covers the imprint of your figurine. Gently tap the pie plate to let any air bubbles in the mixture to escape.
7. Allow to dry thoroughly. (It may dry overnight, or take a couple of days.)
8. Carefully separate the aluminum pie plate from the modeling clay. **Your teacher** may need to cut the plate.
9. Carefully separate the modeling clay from your "fossil".
10. Paint it to add interesting detail.

Teacher Notes

EXPLORING MINERALS

LEARNING INTENTION:
Students will learn about minerals and their properties.

SUCCESS CRITERIA:
- describe different properties of minerals
- recognize the difference between a rock and a mineral
- conduct a study to describe a rock according to its properties
- share findings about one rock with a classmate

MATERIALS NEEDED:
- a copy of "Minerals" Worksheet 1 and 2 for each student
- a copy of "Rock vs. Mineral" Worksheet 3 for each student
- a copy of "A Rock Study" Worksheet 4 for each student
- a collection of different types of minerals (this can be purchased from an educational store/website, or may be borrowed from a school or central library)
- 6 measuring tapes, a weight scale, 2 balance scales, copper coins, nails, pieces of glass, 6 steel files or knives
- magnifying glasses (one per student)
- pencils, chart paper, markers

PROCEDURE:
*This lesson can be done as one long lesson, or be divided into two shorter lessons.

1. Using Worksheet 1 and 2, do a shared reading activity with the students. This will allow for reading practice and learning how to break down word parts in order to read the larger words in the text. Along with the content, discussion of certain vocabulary words would be of benefit for students to fully understand the passage.

Some interesting vocabulary words are:
- luster
- metallic
- surface
- cleavage (to split)
- nonmetallic
- substance

2. Set up mineral exploration centres in the classroom so that groups of students can rotate and use magnifying glasses to get a close look at some mineral samples. Class discussion could follow to allow students to share their findings. Encourage students to comment on luster, color, shape, hardness, etc.

3. Give students Worksheet 3. Read through information as a large group. "Can you guess" section should be done orally. Answers to Worksheet 3:

   (top row left to right) – mineral (fluorite), rock (gneiss), mineral (quartz), mineral (pyrite)
   (bottom row left to right) – rock (sandstone), mineral (amazonite), rock (marble)

4. Give students Worksheet 4. They will choose a rock from the classroom rock collection to study. Once their study is complete, they can share their findings with a friend.

DIFFERENTIATION:
Slower learners may benefit by working as a small group with teacher direction and support in order to provide accurate observations while conducting the rock study. This would result in one record of information, which could be done together, using chart paper and markers. This could be shared with the large group afterwards, if time permits.

For enrichment, faster learners could choose another rock to study.

Minerals

All rocks on Earth contain minerals. Minerals occur in nature and they have a crystal structure and contain chemicals. Minerals have certain characteristics, such as luster, color, cleavage, and hardness. Let's read more about them!

The **luster** of a rock means it may be shiny or dull. The minerals in the rock may be metallic or nonmetallic. Minerals that have a metallic luster make the rock shiny.

Gold is a metallic mineral. It has a shiny surface. Can you spot the gold in this rock?

Talc is a nonmetallic mineral. It has a pearly surface.

The **color** of a mineral depends on the materials that make up the crystals. For example, pure quartz has colorless crystals, but small amounts of other substances can give it a pink or green tint. Amethyst is a purple variety of quartz. Its purple color is caused by the substance iron. The amount of iron in it determines how bright in color it can be.

Cleavage describes how the mineral breaks. When a mineral is cleaved, it may break with one, two, three, or more surfaces. It will break along smooth planes that are close to weak bonding areas within the crystal.

Mica breaks down in one direction. When it is split, it forms thin sheets.

Halite, which is a salt crystal, has cubic cleavage. When it is split, it looks like a cube.

Some minerals like **quartz** do not have cleavage and will only break into uneven pieces with rough surfaces.

The hardness of minerals can be tested by scratching one mineral with another. In 1812, Friedrich Mohs created a scale of hardness. He took ten minerals and numbered them from one, the softest, to ten, the hardest. No mineral on the scale can be scratched by any one that is softer, but it can scratch all that are softer.

Moh's Scale of Hardness

1. **Talc**   (scratch with fingernail)
2. **Gypsum**   (scratch with fingernail)
3. **Calcite** (scratch with copper coin)
4. **Fluorite** (scratch with a steel nail)
5. **Apatite** (scratch with a steel nail)
6. **Feldspar** (scratch with a knife)
7. **Quartz**   (scratches glass)
8. **Topaz**   (scratches glass)
9. **Corundum** (scratches topaz)
10. **Diamond** (scratches another diamond)

Rock vs. Mineral

Are you wondering what the difference is between rocks and minerals? Simply put, **a mineral is made of the same substance throughout. A rock is made up of more than one mineral.**

You have read about and examined some minerals such as feldspar, quartz, and mica. As you looked closely at each, you saw that they were of one substance. Mixed together in different amounts with different types of rocks, they can look very different from their original mineral form. Gneiss, granite, and sandstone are rocks that all have bits of feldspar, quartz, and mica in them.

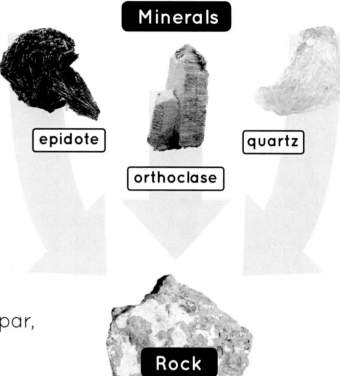

Can You Guess?

Is it a rock or a mineral?

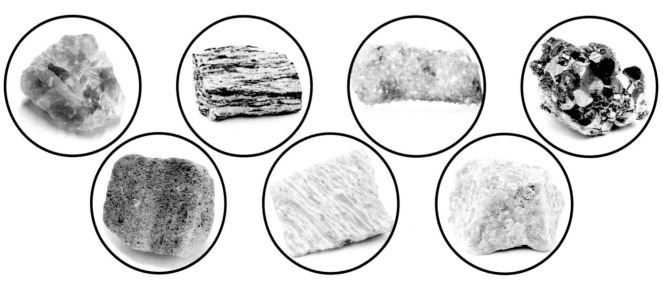

A Rock Study

You have learned a lot about rocks and the minerals in them. It's time to do a rock study!

Choose a rock from your classroom rock collection. Examine its size, weight, shape, color, texture, and hardness. Use these clues to help you determine what type of rock it is and what minerals it may contain.

Characteristics	Qualities about the Rock
Size: small, medium, large? measurement around?	
Weight: put it on the scale, what does it weigh?	
Shape: round, jagged, pointed?	
Color: what colors does it have?	
Luster: shiny, dull, metallic, nonmetallic?	
Texture: rough, bumpy, smooth?	
Hardness: can it be scratched? what scratches it? rate on Moh's Scale?	
Type of rock: igneous? sedimentary? metamorphic?	
Minerals: what minerals are in it?	

Teacher Notes

MINING

LEARNING INTENTION:
Students will learn about the uses of a mineral, the methods of extraction, and the societal and environmental impacts of mining it.

SUCCESS CRITERIA:
- research and complete a written report about a mineral, that includes details about its location, methods used to extract it, the uses of the mineral in our daily lives, the environmental and societal cost of its extraction, and the efforts to reclaim and restore closed extraction sites

MATERIALS NEEDED:
- a copy of "Mining for Information – A Research Project" Worksheet 1, 2, and 3 for each student
- access to computers with internet connection
- clipboards, pencils

PROCEDURE:
*This lesson can be done as one long lesson, or be divided into small research periods.

1. Initiate a class discussion about how minerals are useful to us and where they come from. Give students Worksheets 1, 2, and 3. Explain to them that they will choose a mineral to research. They will look at specific factors that will help them assess the benefits and drawbacks of extracting this mineral.

DIFFERENTIATION:
Slower learners may benefit by working with a partner to research and report on a mineral. Sections of the report could be divided between the partners to lessen the work load.

For enrichment, faster learners could create a poster that details the uses of their mineral to emphasize its importance in our daily lives.

Worksheet 1 Name:

Mining for Information – A Research Project

Choose a mineral that you would like to learn more about. Research this mineral in order to investigate and report on:

- Where is it located?
- What mining method is used to extract it?
- What is this mineral used for in our daily life?
- What is the environmental cost of extracting this mineral?
- What is the cost to society of extracting this mineral?
- How are old mining sites being reclaimed to restore our environment?

Let's Get On Task!

Mineral Name: _____

Mineral Location:

_____
_____
_____
_____
_____

Mining Method Used for Extraction:

_____
_____
_____
_____
_____

Worksheet 2   Name:

A List of the Uses for this Mineral:

- _____
- _____
- _____
- _____
- _____
- _____
- _____
- _____

Illustrate two of the ways that you use this mineral in **your** daily life. Include a caption.

| Worksheet 3 | Name: |

Environmental Cost of Extraction:

_____
_____
_____
_____
_____
_____
_____

Cost to Society for Extraction:

_____
_____
_____
_____
_____
_____
_____

Reclaiming and Restoration of the Land:

_____
_____
_____
_____
_____
_____
_____

Teacher Notes

FUN WITH ROCKS

LEARNING INTENTION:
Students will learn about the presence of carbonates, and the presence of magnetic minerals in certain rocks.

SUCCESS CRITERIA:
- conduct a test to determine the presence of calcite in a sedimentary rock
- record observations in a chart using pictures
- make a conclusion about the type of sedimentary rock
- conduct a test to determine the presence of magnetic minerals in rocks
- research types of minerals in rocks that are magnetic, record findings in a chart

MATERIALS NEEDED:
- a copy of "Testing for Limestone" Worksheet 1 and 2 for each student
- a copy of "What's in that Rock?" Worksheet 3 for each student
- 3 different sedimentary rocks, 3 glass jars, 300 mL of white vinegar, 2 magnifying glasses, a measuring cup (for each pair of students)
- a mixture of rocks (varying in type and color), a strong magnet (for each pair of students)
- masking tape, markers, chart paper, pencils

PROCEDURE:
*This lesson can be done as one long lesson, or be divided into two shorter lessons.

1. Give students Worksheets 1 and 2 and materials to conduct the experiment (ensure that each pair of students has a piece of limestone). Read through the materials needed and what to do section to ensure their understanding of the task. Upon completion of the experiment, students will determine if any of their sedimentary rocks are limestone.

Some interesting facts to know why scientists would want to know determine sedimentary rock as limestone and about the presence of calcite:
- it is used in the construction industry
- calcite is used in antacid tablets to reduce stomach acid
- calcite is used as a whitening agent in paint, and to remove stains in clothing
- ground limestone is sprayed on the walls in coal mines to reduce the dust in the air, it also reflects light in the dark mine

2. Give students Worksheet 3 and the materials to conduct the investigation. Upon completion of the investigation, students will access the internet to research the types of minerals that are present in certain rocks that make them magnetic.

DIFFERENTIATION:
Slower learners may benefit by working as a small group with teacher direction and support in order to provide accurate observations while conducting the experiments. This would result in one record of information, which could be done together, using chart paper and markers. An additional accommodation may be to only have them conduct one of the two experiments.

For enrichment, faster learners could describe some common uses of rocks and minerals, explaining how they are used within the school, at home, or in the community.

Worksheet 1 Name:

Testing for Limestone

You have learned that limestone is a sedimentary rock. Can you tell it apart from other sedimentary rocks? Let's try this simple test to see if a sedimentary rock is limestone!

You'll need:

- 3 different looking sedimentary rocks
- 3 glass jars
- 300 mL of white vinegar
- a magnifying glass
- a measuring cup

What to do:

1. Place a sedimentary rock in each of the glasses.
2. Pour 100 mL of vinegar into the measuring cup.
3. Pour the vinegar over the sedimentary rock in the first glass.
4. Using your magnifying glass, observe what happens in the glass.
5. Record your observations on Worksheet 2.
6. Repeat steps 2 through 5 for each of the remaining rocks.
7. Make a conclusion about which of the rocks may be limestone. Record it on Worksheet 2.

Worksheet 2 Name:

Let's Observe

Rock #1	Rock #2	Rock #3
This is what it looked like when I poured the vinegar over it:	This is what it looked like when I poured the vinegar over it:	This is what it looked like when I poured the vinegar over it:
Did you see any tiny bubbles? _____	Did you see any tiny bubbles? _____	Did you see any tiny bubbles? _____

Bubbling is a sign of a chemical reaction. Vinegar and the mineral **calcite** will create carbon dioxide. If you saw bubbles then the rock is limestone because it contains the mineral calcite.

I ♥ bubble baths!

Let's Conclude

Were any of the rocks you tested limestone? Explain your results. _____

_____

_____

_____

| Worksheet 3 | Name: |

What's in that Rock?

You have seen the chemical reactions that they can cause. Could rocks also be magnetic? Let's investigate this idea!

You'll need:

- an assortment of different types of rocks, both dark colored rocks and light colored rocks
- a strong magnet

What to do:

1. Place the rocks on a table surface, spaced apart.
2. Slowly move the magnet over the top of one of the rocks. If it attracts, it is magnetic.
3. Repeat step 3 using the other rocks.
4. Make some predictions about what could be in the rocks that are magnetic.
5. Use the internet to research what minerals in certain rocks make them magnetic. List your findings in the chart below.

Magnetic Minerals Found in Rocks

_____

_____

_____

_____

_____

_____

_____

EXPERIMENTING WITH EROSION

LEARNING INTENTION:
Students will learn about the effects of wind, water, and ice on our landscape.

SUCCESS CRITERIA:
- make predictions about the effects of wind, water, and ice on rocks and soil
- describe the effect that water has on rocks and soil
- describe the effect that wind has on rocks and soil
- describe the effect that ice has on rocks and soil
- gather data and record observations in a chart using pictures and words
- make conclusions about the effects of water, wind, and ice on our landscape
- make connections about the effects of water, wind, and ice to the environment around us

MATERIALS NEEDED:
- a copy of "Experimenting with Soil Absorption" Worksheet 1, 2, and 3 for each student
- a copy of "Blown Away!" Worksheet 4, 5, and 6 for each student
- a copy of "Shake Down to Break Down!" Worksheet 7 and 8 for each student
- a copy of "The Glacier Effect!" Worksheet 9 and 10 for each student
- samples of soils such as sand, clay, silt, loam, and topsoil (enough for 4 or 5 groups)
- a pan of water, 3 cotton cloths or coffee filters, 3 cups, 3 graduated cylinders, a small garden spade or large spoon (for each group of students)
- an oscillating fan, a table surface, a piece of cardboard (50cm x 100cm)
- 9 sedimentary rocks, 9 igneous or metamorphic rocks, 3 empty coffee cans with lids, 3 clear plastic containers, a large spoon with holes, a sieve, a jug of water (for each group)
- modeling clay, 4 ice cube trays, a jug, 6 spoons, sand with pebbles in it
- access to water, access to a freezer
- pencils, clipboards, chart paper, markers

PROCEDURE:
*This lesson can be done as one long lesson, or be divided into four shorter lessons.

1. Explain to students that they will do an experiment to see what effect water has on different types of soils. Give them Worksheet 1. Read through the question, materials needed, and what to do sections with the students to ensure their understanding. Divide students into pairs or small groups. Give them Worksheets 2 and 3, and the materials to conduct the experiment. Students will make and record observations of the effects water has on each of the soils, then make conclusions and a connection.

2. Explain to students that they will do an experiment to see what effect wind has on different types of soils. Give them Worksheet 4. Read through the question, materials needed, and what to do sections with the students to ensure their understanding. This experiment may be better to do as a whole class, with teacher as operator of the oscillating fan, and a couple of students assigned to hold the cardboard in place. Gather the materials to conduct the experiment. Give students Worksheets 5 and 6. Students will make predictions, make and record observations of the effects wind has on each of the soils, then make conclusions and a connection.

3. Give students Worksheets 7 and 8 and materials to conduct the experiment. Read through the materials needed and what to do section to ensure their understanding of the task. Upon completion of the experiment, students will determine how rocks change form due to erosion caused by moving water.

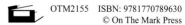

Teacher Notes

4. ***This activity will require preparation ahead of time in order to have ice cubes form.** Give students Worksheet 9. Read about glaciers and how their movement can cause erosion to the landscape. Students will investigate this idea. Give them the materials to conduct the investigation. Read through the 'what to do' section with them to ensure their understanding of this task. Once students have completed the task, they will record observations and make conclusions on Worksheet 10. Students should understand that the sediment (sand and rock) in the glacier is abrasive to the landscape as it moves over it, causing it to erode and become rutted.

*Watching the Magic School Bus episode "Rocks and Rolls" may be beneficial to enhance students' learning about the causes of erosion. Episodes can be accessed at www.youtube.com

DIFFERENTIATION:

Slower learners may benefit by working as a small group with teacher direction and support in order to provide accurate measurements while experimenting with the soils. This would result in one set of data to graph, which could be done together, using chart paper and markers. A further accommodation may be to change the 'making connections' section on Worksheets 3 and 6 to a discussion item within the small group if time permits.

For enrichment, faster learners could research and compile a list of natural occurrences and human activities that cause changes to landscape (e.g., floods, mud slides, avalanches, forest fires, clear cutting in forests, agricultural development, hydroelectric dams). This activity could lead to a large group discussion.

| Worksheet 1 | Name: |

Experimenting with Soil Absorption

Question: What effect does water have on different types of soils?

You'll need:

- a pan of water
- 3 cotton cloths or coffee filters
- 3 cups
- 3 graduated cylinders
- a small garden spade
- soils such as sand, clay, and loam

What to do:

1. Make a prediction to the question and record it on Worksheet 2.

2. Using the gardening spade, scoop some sand and put it on a cotton cloth. Wrap up the cloth. Repeat this step using the clay and then using the loam.

3. Take the cloth wrapped with sand in it, and carefully dip it into the pan of water.

4. Pull the cloth out of the water and let it drip over a cup. Make observations about the amount of water you see dripping from the cloth filled with sand. Draw what you saw in the chart on Worksheet 2.

5. Repeat steps 3 and 4 using the cloth filled with clay, and the cloth filled with loam.

6. Pour the water from each cup into the graduated cylinders. Record the amount of water in each cylinder in the chart on Worksheet 2.

7. Make conclusions and connections about the effect of water on each soil type. Record them on Worksheet 3.

| Worksheet 2 | Name: |

Let's Predict

What effect does water have on different types of soils?

_____

_____

_____

Let's Observe

Cloth filled with sand	**Cloth filled with clay**	**Cloth filled with loam**
This is what it looked like when I pulled the cloth filled with **sand** out of the water:	This is what it looked like when I pulled the cloth filled with **clay** out of the water:	This is what it looked like when I pulled the cloth filled with **loam** out of the water:
When I poured the water from the cup into the graduated cylinder, it measured: _____	When I poured the water from the cup into the graduated cylinder, it measured: _____	When I poured the water from the cup into the graduated cylinder, it measured: _____

Worksheet 3   Name:

Using the data you collected, create a bar graph to show the water absorption of each soil. Be sure to add a title and labels on your graph.

Let's Connect It!

If you wanted to stop a river from overflowing, you could place bags filled with soil along the riverbanks. What kind of soil would you use to fill the bags? Explain your thinking.

_____
_____
_____
_____

| Worksheet 4 | Name: |

Blown Away!

Question: What effect does wind have on different types of soils?

You'll need:

- a fan
- a table surface
- a piece of cardboard (50cm x 100cm)
- 3 graduated cylinders
- a small garden spade
- soils such as sand, clay, and loam

What to do:

1. Make a prediction to the question and record it on Worksheet 5.

2. Fold the cardboard into thirds so that is able to stand upright. Place it at one end of a table that is against a wall.

3. Fill a graduated cylinder with 200 mL of sand.

4. Pour the sand out onto the table.

5. Point the fan at the table, opposite from the cardboard. Turn on the fan for 20 seconds. Make sure that the fan moves from side to side.

6. Make observations about the amount of sand you see left on the table. Draw what you saw in the chart on Worksheet 5.

7. Collect the sand that is left on the table back into the graduated cylinder. How much sand was left on the table? Record this on Worksheet 5.

8. Repeat steps 2 through 6, using the other two types of soils.

9. Make conclusions and connections about the effect of wind on each soil type. Record them on Worksheet 6.

| Worksheet 5 | | Name: | |

Let's Predict

What effect does moving air have on different types of soils?

_____
_____
_____
_____

Let's Observe

Air moving over sand	Air moving over clay	Air moving over loam
This is what it looked like when the fan blew air over the **sand**:	This is what it looked like when the fan blew air over the **clay**:	This is what it looked like when the fan blew air over the **loam**:
When I collected the leftover sand into the graduated cylinder, it measured: _____	When I collected the leftover clay into the graduated cylinder, it measured: _____	When I collected the leftover loam into the graduated cylinder, it measured: _____

| Worksheet 6 | Name: |

Let's Conclude

What effect did the moving air have on the **sand**?

_____
_____
_____

What effect did the moving air have on the **clay**?

_____
_____
_____

What effect did the moving air have on the **loam**?

_____
_____
_____

Let's Connect It!

How could this information be useful in our daily lives?

_____
_____
_____
_____
_____
_____

Worksheet 7	Name:

Shake Down to Break Down!

You have learned that weathering of rocks happens when they are washed away by water or blown away by wind, causing them to be broken into smaller pieces. This is erosion.
Let's experiment with this idea!

You'll need:

- 9 sedimentary rocks
- 9 igneous or metamorphic rocks
- 3 empty coffee cans with lids
- 3 clear plastic containers
- a large spoon with holes, and a seive
- a marker
- a jug of water
- masking tape
- a partner

What to do:

1. Put 3 sedimentary rocks into each of the coffee cans. Then put 3 igneous or metamorphic rocks into each of the coffee cans.

2. Pour some water into each of the coffee cans so that it covers the rocks. Put the lid on each can.

3. Using the masking tape and marker, create a label for the first can that says "no shakes", a label for the second can that says "50" shakes, and a label for the third can that says "500 shakes". Do this same step for each of the clear plastic containers.

4. Shake the can labeled "50 shakes", 50 times. Shake the can labeled "500" shakes, 500 times. (Take turns).

5. Remove the rocks from each container, keeping the piles separate.

6. Pour the water from each can through the sieve, into its matching labeled plastic container. Put the remaining rocks in the sieve in its matching pile.

7. Record your observations and conclusions about the rocks and water on Worksheet 4.

| Worksheet 8 | Name: |

Let's Observe

"No Shakes"	"50 Shakes"	"100 Shakes"
This is what **the rocks** looked like when I poured the water through the sieve:	This is what **the rocks** looked like when I poured the water through the sieve:	This is what **the rocks** looked like when I poured the water through the sieve:
Describe what **the water** looked like once it had passed through the sieve. _____ _____ _____ _____	Describe what **the water** looked like once it had passed through the sieve. _____ _____ _____ _____	Describe what **the water** looked like once it had passed through the sieve. _____ _____ _____ _____

Let's Conclude

What caused the rocks to change form?

_____

_____

Use your results to explain what happens to rocks that are carried down the river by water current.

_____

_____

_____

The Glacier Effect!

A glacier is a mass of compacted snow and ice that forms in very cold climates. Glaciers that have significant weight will move with the help of gravity. When a glacier moves, it scrapes sediment and rocks that are frozen to surfaces over which they flow. This abrasive action is **erosion!**

Let's try an experiment to examine the effect that ice has on our landscape!

You'll need:

- modeling clay
- an ice cube tray
- a jug of water
- a cup of sand mixed with tiny pebbles
- access to water

What to do:

1. Put some sand mixture into the bottom of each section of the ice cube tray.

2. Fill the tray up with water. Place it in the freezer until frozen solid.

3. Take a bit of modeling clay and flatten it out. (This will represent our landscape.)

4. Take out a frozen, sand-filled ice cube from the tray. (This will represent a glacier that has picked up sediment.)

5. Rub it over top of the modeling clay to simulate the movement of ice as it melts over top of our landscape.

6. Make and record your observations on Worksheet 10.

7. Make and record conclusions about the effects of ice on our landscape.

| Worksheet 10 | Name: |

Let's Observe

This is what the landscape looked like *before* it was eroded by melting ice mixed with sediment:	This is what the landscape looked like *after* it was eroded by melting ice mixed with sediment:
Description of the land: _____ _____ _____ _____	Description of the land: _____ _____ _____ _____

Let's Conclude

Use your observations to explain how our landscape changes as we experience changes in temperature.

_____
_____
_____
_____

Teacher Notes

EROSION PREVENTION!

LEARNING INTENTION:
Students will learn about the signs of erosion and the techniques to prevent it from occurring.

SUCCESS CRITERIA:
- discover and record signs of erosion in the neighborhood
- create a system to control soil erosion
- describe the effect that moving water has on soil
- describe the impact vegetation has on erosion prevention
- determine methods to prevent erosion

MATERIALS NEEDED:
- a copy of "Signs of Erosion" Worksheet 1 for each student
- a copy of "Controlling Erosion" Worksheet 2, 3, and 4 for each student
- 2 large plastic bins (deep and rectangular), soil to fill the bins, a piece of grass sod, an empty milk container (cardboard), an empty jug, 4 litres of water (for each pair of students)
- a sharp knife or scissors
- pencils, clipboards

PROCEDURE:
*This lesson can be done as one long lesson, or be divided into two shorter lessons.

1. Give students Worksheet 1, a clipboard, and a pencil. Take them out on a walk through the neighborhood to look for signs of soil erosion. (Areas to look for may be road shoulders, along a stream bank, under an eavestrough spout or sprinkler.) Upon completion of Worksheet 1, have students share their findings with the large group. Engage students in a discussion about the dangerous/ negative effects that moving water has on our landscape.

2. Explain to students that they will do an experiment to see what effect plant life has on soil erosion. Give them Worksheet 2. Read through the question, materials needed, and what to do sections with the students to ensure their understanding of the task. (To save time, have milk containers available that already have holes in the bottom of them.) Divide students into pairs or small groups. Give them Worksheets 3 and 4, and the materials to conduct the experiment. Students will make and record observations of the effect that grass has on preventing soil erosion. **(Grasses have fibrous roots that spread out into the soil in many directions. This helps to hold the soil together, so water does not easily erode it.)** Students will make a conclusion then work in pairs to make connections between human activity/efforts and soil erosion. An option is to come back together as a large group to share their responses. A question to pose: What erosion prevention measures have you seen in your neighborhood? Where are they located?

DIFFERENTIATION:
Slower learners may benefit by working as a small group with teacher direction in order to construct a system to control soil erosion. This would allow for support to be given in the description of the effect that water has on soil and the impact vegetation has on controlling erosion. The 'think-pair-share' activity could be done as a small group on chart paper.

For enrichment, faster learners could access the internet to research techniques for controlling erosion (e.g., mulching, wind breaks, terracing, sediment basins, and crop rotation). An option is to have these learners present their findings to the large group.

Worksheet 1 Name:

Signs of Erosion

Take a walk in your neighborhood. Look for signs of soil erosion. Illustrate four signs of soil erosion. Describe how the erosion happened.

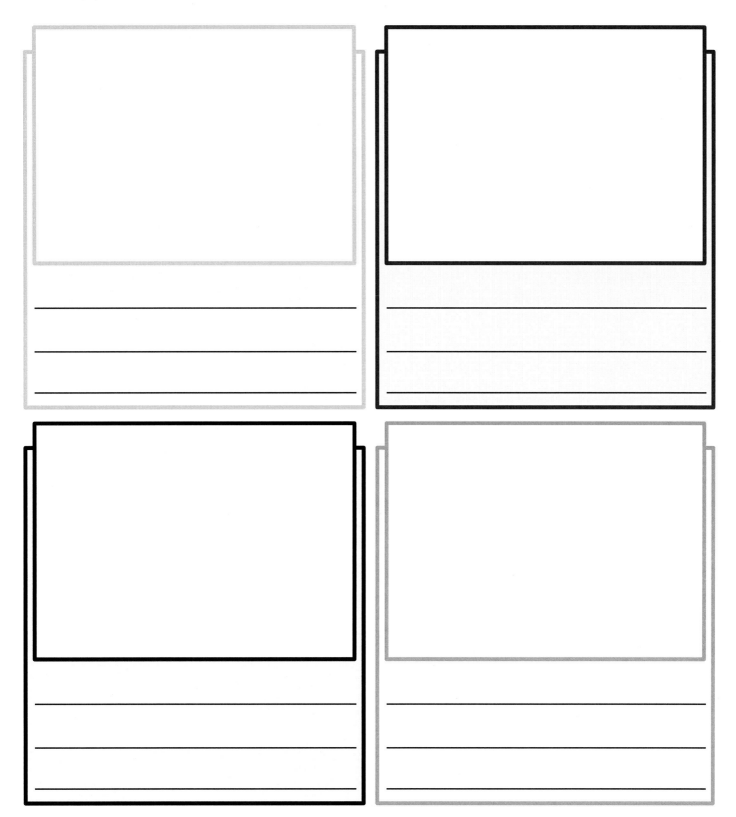

| Worksheet 2 | Name: |

Controlling Erosion

Erosion...you know what it is, you know about the impact it can have on our landscape, now what can you do about it?

Question: Do grass and plants help to prevent erosion?

You'll need:

- 2 large plastic bins (deep and rectangular)
- an empty milk container (cardboard)
- a piece of grass sod
- a sharp knife or scissors
- soil
- a partner
- a garden shovel
- an empty jug
- 4 litres of water

What to do:

1. Make a prediction and record it on Worksheet 3.

2. Using the garden shovel, fill both of the large plastic bins with some soil. Shape it to form one side of a hill.

3. In one of the bins, place the piece of grass sod onto the soil mound. This will represent a grassy hillside.

4. **Your teacher** will make holes in the bottom of the milk container.

5. Your partner will hold the milk container over the bin of only soil.

6. You will use the jug to pour 2 litres of water in through the milk container, while your partner moves it over top of the hill of soil. This will represent rain.

7. Repeat steps 5 and 6, using the bin of soil with grass sod.

8. Observe the amount of water that collects at the bottom of each hill, and the consistency and color of the water. Record your observations.

9. Make conclusions and connections about soil erosion on Worksheet 4.

| Worksheet 3 | Name: |

Let's Predict

Do grass and plants help to prevent erosion? Explain your thinking. _____

_____

_____

Let's Observe

Bin with **soil** hillside	Bin with **grassy** hillside
Illustration of the **soil** hillside after it rained:	Illustration of the **grassy** hillside after it rained:
Describe the water collection at the bottom of the hill after it had rained. _____ _____ _____ _____ _____	Describe the water collection at the bottom of the hill after it had rained. _____ _____ _____ _____ _____

| Worksheet 4 | Name: |

Let's Conclude

Do grass and plants help to prevent erosion? Explain your thinking. _____
_____
_____

Let's Connect It!

Think | Pair | Share

With a partner, do some thinking and sharing of ideas about things that cause soil erosion.

Complete the web below by adding things that humans can do **to prevent** our landscape from eroding.

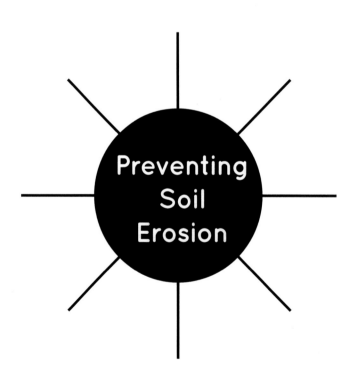

Teacher Notes

WEATHER PATTERNS

LEARNING INTENTION:
Students will learn about different types of weather and daily weather patterns.

SUCCESS CRITERIA:
- discuss different types of weather that is experienced on the Earth
- provide written descriptions of different types of weather
- research and record daily weather patterns
- determine temperature readings and make connections to weather
- take daily temperature readings, record and compare them to predicted daily highs

MATERIALS NEEDED:
- a copy of "What is Weather?" Worksheet 1 and 2 for each student
- a copy of "Going on a Weather Watch!" Worksheet 3 for each student
- a copy of "Taking a Reading" Worksheet 4 and 5 for each student
- a copy of "What's the Temperature?" Worksheet 6 for each student
- a large thermometer (teaching thermometer)
- thermometers (one per student)
- daily access to the Internet, radio, television, or newspapers
- chart paper, markers
- pencils, clipboards

PROCEDURE:
*This lesson can be done as one long lesson, or be divided into two shorter lessons. Items 1 and 3 can be done at school. Items 2 and 4 contain a homework component.

1. Divide students into partners. Give students Worksheet 1. With their partner, students will brainstorm and discuss types of weather that they have seen or heard of happening on Earth. Once students have had time to discuss and record their ideas, an option would be to come together as a large group to discuss their answers. Recording answers on a chart paper would be beneficial for future reference. Give students Worksheet 2 to complete.

2. Explain to students that they will do some weather watching for one week. Each day, they will research daily weather patterns by any such method as the internet, watching a weather station on television, listening to radio broadcasts, or reading the newspaper. Using Worksheet 3, students will record daily temperatures, wind chill / humidex values, wind speed and direction, and describe cloud cover trends of the day. They will also record precipitation type, amounts, and the probability of occurrence. A class discussion about the meaning of these weather terms would be beneficial. (*An option would be to track daily weather patterns as a large group, so that the students are recording the same information. Tracking this data on a large chart would allow for reference for the activity explained in item 4, which will be conducted simultaneously.*)

3. Using a large (teaching) thermometer, teach students how to read the temperature on a thermometer. Ensure students are able to relate a certain temperature to a type of weather. Give students Worksheets 4 and 5 to complete.

4. Give students Worksheet 6 and a thermometer. They will take daily temperature readings outdoors and compare their results each day with the predicted daily high temperature that they researched and recorded on Worksheet 3. *Daily time for readings can be adjusted to another **afternoon** time, depending on the schedule of your school day.

Teacher Notes

DIFFERENTIATION:

Slower learners may benefit by working with a faster learner in order to complete Worksheet 1. This would allow for peer support in recording ideas and weather words. Eliminating the written output expectation on Worksheet 2 would be a further accommodation. These learners may also benefit from working in a small group to read and record daily temperatures.

For enrichment, faster learners could create a double bar graph to display the daily high and low temperatures they recorded on Worksheet 3. Graphs should include a title, labels, appropriate scales, and concluding statements. A further step would be to have these learners determine a mode and a median for their data.

| Worksheet 1 | Name: |

What is Weather?

Think | **Pair** | **Share**

With a partner, do some thinking and sharing of ideas about the different types of weather you have seen or heard about happening on our planet.

In the box below, make a list of the weather words that you use in your discussion.

Worksheet 2   Name:

Illustrate and describe some types of weather that you have seen. Use your weather words on Worksheet 1 to help you.

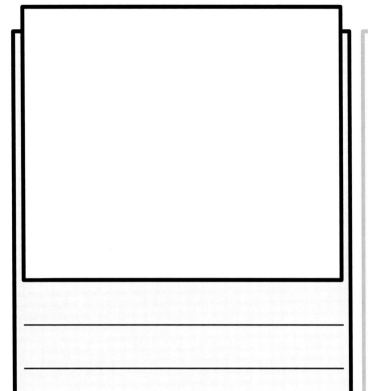

Going on a Weather Watch!

You have talked about the weather, now it is time to look more closely at the daily weather patterns in your area. Let's get tracking!

	Sunday	Monday	Tuesday	Wednesday	Thursday	Friday	Saturday
Temperature	high ____ ° low ____ °	high ____ ° low ____ °	high ____ ° low ____ °	high ____ ° low ____ °	high ____ ° low ____ °	high ____ ° low ____ °	high ____ ° low ____ °
Wind chill/ humidex reading	feels like ____ °	feels like ____ °	feels like ____ °	feels like ____ °	feels like ____ °	feels like ____ °	feels like ____ °
Cloud cover type/sunny periods							
Precipitation	Type: Amount:	Type: Amount:	Type: Amount:	Type: Amount:	Type: Amount:	Type: Amount:	Type: Amount:
Probability of Percipitation	%	%	%	%	%	%	%
Wind speed/ direction							

Worksheet 4   Name:

Taking a Reading

A thermometer measures the temperature. It helps us to know how hot, how warm, or how cold things are.

We can use a thermometer to measure the temperature outside.

Let's take some readings! Record the temperature shown on each thermometer, and circle Celcius (°C) or Fahrenheit (°F).

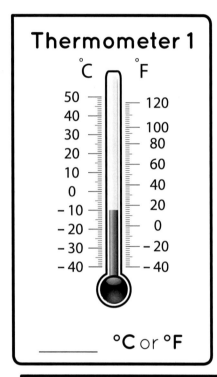

Thermometer 1

_____ °C or °F

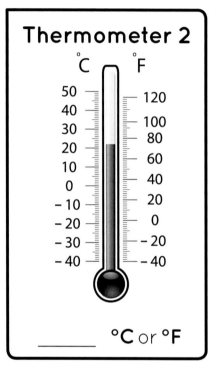

Thermometer 2

_____ °C or °F

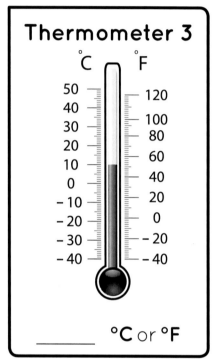

Thermometer 3

_____ °C or °F

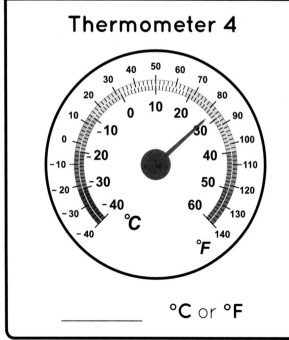

Thermometer 4

_____ °C or °F

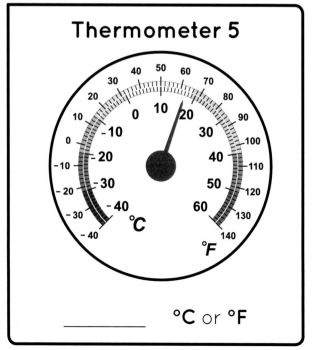

Thermometer 5

_____ °C or °F

| Worksheet 5 | Name: |

Use **"warmer than"** or **"cooler than"** to compare the temperatures on the thermometers on Worksheet 4.

Thermometer 3 is _____ **Thermometer 1**.

Thermometer 3 is _____ **Thermometer 2**.

Thermometer 4 is _____ **Thermometer 5**.

Thermometer 2 is _____ **Thermometer 5**.

Which thermometer on Worksheet 4 could be used to describe the temperature in each picture below?

Thermometer ___

Thermometer ___

Thermometer ___

Worksheet 6 Name:

What's the Temperature?

Now that you are familiar with reading a thermometer, take a temperature reading on each day of the week.

Refer back to the information you collected on Worksheet 3 to determine if the daily predicted high has been reached.

Let's Observe

	Temperature reading at 2:00 pm	Daily predicted high temperature	Was the predicted daily high reached?
Sunday			
Monday			
Tuesday			
Wednesday			
Thursday			
Friday			
Saturday			

PRECIPITATION

Teacher Notes

LEARNING INTENTION:
Students will learn about the water cycle and measure weather in terms of precipitation.

SUCCESS CRITERIA:
- describe how the water cycle creates precipitation in our natural environment
- recreate the water cycle
- construct a rain/ snow gauge to measure precipitation amounts
- collect data and display results in a graph
- make conclusions and connections to the environment

MATERIALS NEEDED:
- a copy of "Where Does Precipitation Come From?" Worksheet 1 for each student
- a copy of "Exploring the Water Cycle" Worksheet 2 and 3 for each student
- a copy of "A Measure of Precipitation!" Worksheet 4, 5, and 6 for each student
- a clear glass jar, a foil pan, about 6 ice cubes (a set for each group of students)
- a large jug of water, a few desk lamps, a large funnel
- 2 tall clear glass jars (or beakers) that are about 30 cm in height, a ruler, a permanent marker, some plasticine, some rocks (a set for each group of students)
- pencils, clipboards, chart paper, markers

PROCEDURE:
*This lesson can be done as one long lesson, or be divided into two shorter lessons.

1. Give students Worksheet 1. Read through the information about nature's water cycle. Discuss the concepts of 'condensation', 'precipitation', and 'evaporation' with students to ensure their understanding of how each is created and part of the cycle.

2. Explain to students that they will recreate the water cycle in order to examine it more closely. Divide students into groups and give them Worksheets 2 and 3, and the materials to conduct the examination. Read through the materials needed and what to do sections with students to ensure their understanding of the task. Students will record their observations and conclusion on Worksheet 3.

3. **This activity can be done in small groups or as one large group, and *it will span over a period of 5 days.*** Give students Worksheets 4, 5, and 6, and the materials to create the rain (or snow) gauge. In order for students to complete the graph on Worksheet 6, they will need to access the internet to research previous daily precipitation amounts in their area. An option is to graph these findings as a large group on chart paper.

DIFFERENTIATION:
Slower learners may benefit by working as a small group with teacher direction to discuss their conclusions on Worksheet 3, instead of writing them out. Also, if the rain gauge activity is not done as a large group, it would be beneficial for these learners to work on it as small group with teacher support in order to take accurate daily measurements. The graphing activity could be done as a small group on chart paper.

For enrichment, faster learners could access the internet to research the average daily precipitation amounts for the month you are in, for your area, for the past 10 years. They can display these results in a bar graph. They can make a conclusion about typical climate conditions for the area for the month you are in. An option is to have these learners present their findings to the large group.

Where Does Precipitation Come From?

Precipitation is the moisture that falls from the sky. Precipitation can be in different forms. It can fall in forms such as rain, snow, sleet, or hail. Have you ever wondered where precipitation comes from exactly? Let's learn how precipitation is created in our environment.

Condensation happens when water in a gas form, meets cooler air. It will form a cloud in the sky where this gas changes into droplets of water.

These droplets of water, called precipitation, fall back down to Earth. **Precipitation** can be in the form of rain, snow, sleet, or hail. It wets the ground, and fills up rivers, lakes, oceans, and streams.

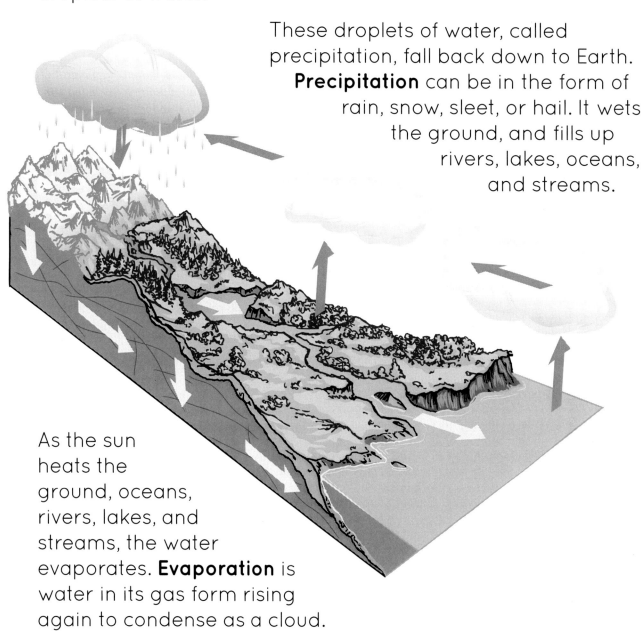

As the sun heats the ground, oceans, rivers, lakes, and streams, the water evaporates. **Evaporation** is water in its gas form rising again to condense as a cloud.

Worksheet 2　　Name:

Exploring the Water Cycle

You know where precipitation comes from, now you can create your own water cycle!

You'll need:

- a clear glass jar, half filled with water
- a foil pan
- about 6 ice cubes
- a desk lamp

What to do:

1. Place the foil pan on top of the glass jar that is half filled with water.

2. Put the ice cubes on the foil pan.

3. Shine the light from the lamp directly at the ice cubes.

4. Make observations and record them on Worksheet 3.

5. Make conclusions and record them on Worksheet 3.

| Worksheet 3 | Name: |

Let's Observe

Describe what you see sticking to the bottom of the aluminum pan that is inside the glass.

_____

_____

_____

If the water in the bottom of the jar represents a lake, then:

What do you think the lamp represents?

_____

What do you think the aluminum pan and the ice cubes represent?

_____

_____

Let's Conclude

Use your observations to explain what is happening inside the jar.

_____

_____

_____

_____

Challenge question:

Do you think the water in the jar will ever get used up? Explain your thinking.

_____

_____

_____

_____

A Measure of Precipitation!

Have you wondered how precipitation is measured? Rain can be collected and measured by using a **rain gauge**. A rain gauge measures rainfall in millimetres (mm).

Snowfall is another form of precipitation that can be collected and measured. Snowfall is measured in centimetres (cm). Did you know that 10 cm of fallen snow is equivalent to 10 mm of rainfall?

Let's investigate precipitation a little further by making a rain gauge or a snow gauge to measure just how much of it falls from the sky!

You'll need:

- plasticine
- a ruler
- a permanent marker
- some rocks
- a tall clear glass jar (two jars if you are collecting snow)
- a funnel with a mouth wider than the glass jar

What to do:

(to measure rain)

1. Using the ruler and marker, measure and mark off each 2 mm going up the glass until you reach 20 mm.

2. Use the plasticine to make a ring around the top of the jar.

3. Place the funnel over the top of the glass, so that it is sitting on the ring of plasticine and secured.

4. Place the rain gauge outside in a clear area. Surround the bottom of the glass with rocks so that it does not tip over.

5. Measure the height of water in the gauge **at the same time every day**, for a period of 5 days.

6. Record the daily rainfall amounts on Worksheet 6.

7. Record observations and make conclusions about precipitation patterns in your area on Worksheet 6.

What to do:

(to measure snow)

1. Using the ruler and marker, measure and mark off each 2 cm going up both of the glasses until you reach 20 cm on each one.

2. Place the snow gauge (one jar, with no funnel) outside in a clear area. Surround the bottom of the glass with rocks so that it does not tip over.

3. Measure the height of snow in the gauge **at the same time every day**, for a period of 5 days. Do this by:
    - replacing the measured glass with the empty one
    - take the measured glass indoors so that the snow melts completely
    - then measure the height of the water

4. Record the daily snowfall amounts on Worksheet 6.

5. Record observations and make conclusions about precipitation patterns in your area on Worksheet 6.

Worksheet 6	Name:

Precipitation Amounts

	Day 1	Day 2	Day 3	Day 4	Day 5
Rainwater (mm)					
Snow (cm)					
Melted Snow (mm)					

Research the precipitation amounts in your area for all 20 days *before* you started measuring. Plot these amounts *along with your results* using a bar graph. Remember to label your graph!

Explain the type of climate you have experienced this past month. _____
_____
_____

Teacher Notes

WEATHER INSTRUMENTS

LEARNING INTENTION:
Students will learn about the measure of weather in terms of wind speed and direction.

SUCCESS CRITERIA:
- construct a wind vane to measure wind direction
- observe wind direction and record it in a diagram
- construct an anemometer to measure wind speed
- determine wind speed through observation of RPM and calculations
- determine wind chill factors from a set of given data

MATERIALS NEEDED:
- a copy of "Which Way is the Wind Blowing?" Worksheet 1 and 2 for each student
- a copy of "Measuring Wind Speed" Worksheet 3, 4, and 5 for each student
- a copy of "The Chill Factor!" Worksheet 6 and 7 for each student
- a copy of "In Search of the Weather!" Worksheet 8 for each student
- 2 paper plates, a tall pencil with an eraser at the end of it, a straw, 2 pieces of card stock, a marker, a lump of modeling clay, a straight pin, a compass (a set for each group of students)
- 2 straws, a straight pin, a tall pencil with an eraser at the end of it, 5 small Styrofoam cups, a stop watch, a calculator, (a set for each group of students)
- 2 or 3 fans
- masking tape, a few staplers, a few single hole punchers, scissors, rulers
- pencils, markers, clipboards

PROCEDURE:
*This lesson can be done as one long lesson, or be divided into three shorter lessons.

1. Explain to students the purpose of a wind vane. Students will have an opportunity to create one of their own. Give them Worksheets 1 and 2, and the materials needed. Read through the materials needed and what to do sections with students to ensure their understanding of the task. Students can work individually, or be divided into groups to complete this task.

2. Explain to students the purpose of an anemometer. Students will create one of their own. Divide students into pairs and give them Worksheets 3, 4, and 5, and the materials needed. Read through the materials needed and what to do sections with students to ensure their understanding of the task. *Students may need a lesson on what **circumference of a circle** means and what **RPM** (revolutions per minute) means before beginning this task.*

3. Discuss with students the concept that wind speed can affect the feel of the air temperature, thus known as the wind chill factor. Give students Worksheets 6 and 7 to complete.

4. Give each student the word search puzzle on Worksheet 8 to complete.

DIFFERENTIATION:
Slower learners may benefit by working as a small group with teacher direction, using one anemometer to measure RPM, then calculating the wind speed together. The calculations could be recorded on chart paper and later shared with the large group as an example of how to wind speed is determined.

For enrichment, faster learners could access different medias such as radio, newspaper, internet to gather data on the daily weather forecast. They can use this to create and perform their own detailed weather forecast for the large group.

| Worksheet 1 | Name: |

Which Way is the Wind Blowing?

Wind is the result of air in motion. Air moves from high pressure areas to low pressure areas. The strength of the wind is related to the differences in these pressure areas.

A **wind vane** is an instrument that is used to measure wind direction. Create a wind vane of your own to measure which way the wind is blowing!

You'll need:

- 2 paper plates
- a straw
- a marker
- a ruler
- scissors
- a compass
- a tall pencil with an eraser at the end of it
- 2 pieces of card stock (10cm x 10cm)
- a lump of modeling clay
- a straight pin
- masking tape
- a windy day

What to do:

1. Invert one of the paper plates and write **N** at the top, write **S** at the bottom, write **E** at the right side, and write **W** at the left side of the plate.

2. Use the pencil to poke a hole through the centre of this plate.

3. Place the other paper plate on a flat surface. Put the lump of modeling clay in the centre of the plate.

4. Stick the tip of the pencil into the modeling clay. Pull down the top plate so that it meets the bottom plate. Tape them together.

5. Place the straw horizontally on the eraser of the pencil so that is extended further at one end. **Your teacher** will push the stick pin through the straw and into the eraser of the pencil.

| Worksheet 2 | Name: |

6. From the card stock, cut a rectangle (8cm x 4cm) and a triangle (6cm x 6cm x 6cm). This will be the wind vane's tail and arrow.

7. Tape the tail to the longer end of the straw and tape the arrow to the shorter end of the straw.

8. Place a compass on a flat surface in an open area outdoors. Place your wind vane on the same flat surface so that the **N** on the plate matches the **N** direction that the compass indicates.

9. Watch the straw rotate around the pin. Observe which direction the wind is blowing from. (The arrow will point in the direction the wind is blowing from). Record your observations.

Let's Observe

Record your observations by drawing a detailed diagram of your wind vane showing which direction the wind is blowing from.

The wind is blowing from the _____

Measuring Wind Speed

An **anemometer** is a weather instrument that is used to measure wind speed. Create an anemometer of your own to measure just how fast the wind is blowing!

You'll need:

- a straight pin
- 2 straws
- a marker
- a ruler
- scissors
- a stop watch
- a tall pencil with an eraser at the end of it
- 5 small Styrofoam cups
- a single hole punch
- a stapler
- a fan
- a calculator

What to do:

1. Use the marker to color one of the Styrofoam cups.

2. Using the single hole puncher, punch a hole into each of the 4 remaining cups, about 2 cm below the rim.

3. Take the colored cup and punch 2 holes directly opposite of each other, about 2 cm below the rim. Then, punch two more holes directly opposite each other, about 1 cm below the rim.

4. Use the push pin to make a hole in the centre of the bottom of the colored cup. Then use the pencil to enlarge the hole so that the pencil fits through. Remove the pencil for now.

| Worksheet 4 | Name: |

5. Put a straw into the hole of one of the cups that only has one hole. It should go in only about 1 cm. Bend the end of the straw that is inside the cup and staple it down. Feed the other end of the straw through the 2 holes in the colored cup and feed it into the hole of one of the other cups. Bend this end of the straw that is inside this cup and staple it down. *Make sure the cups are facing in opposite directions!*

6. Repeat step 5 using the other straw and remaining cups. *Make sure that the open end of each cup faces the bottom of the cup in front of it.*

7. Insert the eraser end of the pencil into the bottom of the colored cup.

8. **Your teacher** will carefully push the straight pin through the two straws and into the eraser.

9. Turn on the fan. Hold your anemometer in front of it. One partner will count how many turns the colored cup makes in one minute **(RPM)**, while the other partner starts and stops the stopwatch.

10. Use the RPM number to calculate the wind speed in the chart on Worksheet 5.

| Worksheet 5 | Name: |

Let's Calculate

Use a calculator to help you calculate the wind speed.

1) The number of **RPM** the anemometer made was _____.

2) The diameter of the anemometer is _____ cm.

3) The diameter **X** 3.14 = _____ cm. This is the value of the circumference of the anemometer.

4) The **RPM X** the circumference = _____ cm.

5) Convert your answer in cm to km:
_____ cm = _____ km

6) My anemometer calculated a wind speed of:
_____ km per minute

7) **Wind speed is typically reported in km per hour.** You know that there are 60 minutes in 1 hour. Multiply your answer from step #6 by 60. Record the answer in the blank below, and in the digital anamometer screen.

My anemometer calculated a wind speed of:

_____ **km per hour**

The Chill Factor!

Very often on a cold blustery winter's day, we hear the weather forecaster mention the **wind chill factor**. Wind chill happens when the air temperature and the wind combine to make the air feel colder than the actual temperature reading on the thermometer outside.

Let's Examine

Air Temperature (°C)	Wind Speed in km per hour				
	10km	20km	30km	40km	50km
0°C	-2°C	-7°C	-11°C	-13°C	-15°C
-5°C	-7°C	-13°C	-17°C	-20°C	-22°C
-10°C	-12°C	-19°C	-24°C	-27°C	-29°C
-15°C	-17°C	-25°C	-31°C	-34°C	-36°C
-20°C	-22°C	-31°C	-37°C	-41°C	-44°C
-25°C	-27°C	-37°C	-44°C	-48°C	-51°C

1. If the air temperature was -10°C and the wind speed was 30 km per hour, what would the wind chill temperature be?

   _____

2. If the air temperature was -5°C, and the wind speed was 20 km per hour, what would the wind chill temperature be?

   _____

3. If the wind chill temperature was -41°C, what would the air temperature *and* the wind speed be?

   _____

Worksheet 7   Name:

4. What day would be colder?

   a) a day when the air temperature is -15°C and the wind speed is 10 km per hour

   or

   b) a day when the air temperature is -5°C and the wind speed is 50 km per hour

Use the information in the chart on Worksheet 6 to explain your answer.

_____
_____
_____
_____

5. Study the information in the chart on Worksheet 6. Tell what the air temperature and the wind speed would be on:

   a) the coldest possible day

   _____
   _____
   _____
   _____
   _____

   b) the warmest possible day

   _____
   _____
   _____
   _____

Worksheet 8   Name:

In Search of the Weather!

Look for these weather words in the word search puzzle below.

Precipitation	Forecast	Hot	Frost
Sunshine	Cold	Evaporation	Rain
Windy	Thermometer	Anemometer	Snow
Cloudy	Condensation	Cool	Thunder
Temperature	Warm	Climate	Lightning

```
C O N D E N S A T I O N N
T O T G W O N S D Z E O T
T E O O N Y M L D V I T H
G N M L H I O V A T R S E
T I P P C C N P A O A O R
H H X L E L O T R Z I R M
U S V C K R I Y H Y N F O
N N X P A P A M D G H L M
D U F T I K R T A N I A E
E S I C W A R M U T I L T
R O E Y D U O L C R E W E
N R F O R E C A S T E P R
P R E T E M O M E N A I A
```

WASTE

LEARNING INTENTION:
Students will learn about the waste that is a result of human activity versus plant and animal waste, and the ways waste is managed in the natural world.

SUCCESS CRITERIA:
- define the meaning of waste and provide examples of it on our planet
- determine which waste is due to human activity vs. plant and animal waste
- describe how waste is managed in the natural world
- create an earthworm farm
- record observations about earthworm activity
- make a conclusion about the earthworm's role in waste management

MATERIALS NEEDED:
- ask each student to bring in a large wide-mouthed glass jar with a lid
- a copy of "What is Waste?" Worksheet 1 for each student
- a copy of "Human Activity vs. Nature" Worksheet 2 and 3 for each student
- a copy of "A Tree in a Forest" Worksheet 4 for each student
- a copy of "The Earthworms Clean Up!" Worksheet 5 and 6 for each student
- dictionaries or access to the internet
- read aloud about plant and animal waste management (see suggestion in #4 of procedure section)
- soil such as sand and loam or topsoil (enough to fill large jars for each student)
- a hammer and a nail, masking tape, a jug of water, a few small cups
- earthworms (2 or 3 per student)
- vegetable or fruit scraps
- large sheets of black construction paper (2 per student)
- pencils, markers, chart paper, clipboards
- assorted paint colors, paint brushes, and pieces of white art paper (optional)

PROCEDURE:
*This lesson can be done as one long lesson, or be divided into four or five shorter lessons.

1. Explain to students that they will learn about waste. Give them Worksheet 1 and a dictionary, or allow them access to the internet. They will research the meaning of waste, then engage in a "Think-Pair-Share" activity with a partner to brainstorm some examples of waste they have seen in our world. A follow up option is to come back as a large group and record their ideas on chart paper that could be used for a later activity.

2. Give students Worksheet 2 to sort their ideas of waste into two categories, these being waste caused by human activity and waste caused by plants and animals. Give students a clipboard to put Worksheet 2 on, and a pencil. Take them on a walk through the neighborhood to look for signs of waste. They will add these to Worksheet 2.

3. Give students Worksheet 3 to complete. Once again they will engage in a "Think-Pair-Share" activity with a partner to brainstorm some examples of how the natural world manages its own waste. A follow up option is to come back as a large group and record their ideas on chart paper that could be used for a later activity.

4. Read *A Tree in a Forest* (Author: Jan Thornhill) to the students. Ask students to listen for examples of how nature manages its own waste. Record students' ideas on chart paper as you read through the story. Give students Worksheet 4 to complete.

Teacher Notes

5. Students will have an opportunity to create their own earthworm farms. Give them Worksheets 5 and 6, and the materials to create the farms. Read through the question, materials needed, and what to do sections on Worksheet 5 with the students to ensure their understanding of the task. Students will make and record observations of the earthworm farms as they are created and again 24 hours later. They will make a conclusion about the purpose earthworms have in nature's waste management system.

*As an activity to enhance the learning about the necessity of decomposition in the natural world, show students The Magic School Bus episode called "Meets the Rot Squad". Episodes can be accessed at www.youtube.com.

DIFFERENTIATION:

Slower learners may benefit by partnering up with a faster learner to complete the "Think-Pair-Share" activities. Also, this type of pairing could be of benefit while these learners complete Worksheet 2.

For enrichment, faster learners could choose one of their illustrations on Worksheet 4 and re-create it on a larger paper. They can paint this and display it on a bulletin board.

| Worksheet 1 | Name: |

What is Waste?

What exactly is "**waste**"? Use a dictionary to find the definition of this word, or use the internet to research its meaning. Record your answer below.

Waste is _____

_____

_____

Think **Pair** **Share**

With a partner, do some thinking and sharing of ideas about waste you have seen or heard about happening on our planet.

In the box below, record your ideas of waste.

Worksheet 2　　　Name:

Human Activity vs. Nature

Look at the ideas of waste that you recorded on Worksheet 1. Which ones are caused by human activity? Which ones are caused by plants and animals in the natural world?

Sort It Out!

Using point form, sort your ideas in the chart below.

Waste caused by human activity	Waste caused by plants and animals

| Worksheet 3 | Name: |

1) Take a walk in your neighborhood. Look for signs of waste. Add them to your sorting chart on Worksheet 2.

2) Look back at your sorting chart. What produces more waste, human activity or plant and animals? Justify your answer.

_____
_____
_____
_____
_____

Think   Pair   Share

With a partner do some thinking and sharing of ideas of how the plant and animal world manages its waste. In the box below, record your ideas.

| Worksheet 4 | Name: |

A Tree in a Forest

Illustrate and describe some ways that nature managed its own waste in the story *A Tree in a Forest*.

| Worksheet 5 | Name: |

The Earthworms Clean Up!

When an earthworm eats and then digests food such as grass, leaves, or other vegetation, its castings add nutrients back to the soil that plants need in order to grow. Some people make worm farms in order to make rich soil to add to their gardens at home. Let's give this a try!

You'll need:

- a large jar with a lid
- masking tape
- 2 or 3 earthworms
- a hammer and a nail
- soil (a mix of loam and sand)
- vegetable or fruit scraps
- a small cup of water
- 2 large sheets of black construction paper

What to do:

1. Fill the jar until it is three-quarters full of loam soil. Add a thin layer of sand to the top of the soil.

2. Throw some vegetable or fruit scraps on top of the sand.

3. Moisten the soil with a little bit of water. Then add the earthworms.

4. Using the hammer and a nail, **your teacher** will make some holes in the lid of the jar. Put the lid on the jar to close it.

5. On Worksheet 6, draw your observations of the jar and its contents.

6. Cover the sides of the jar with black construction paper so that no light can get in. **Earthworms like it dark!**

7. Leave the jar for a day.

8. The next day, remove the paper. Observe the changes in the jar. Record them on Worksheet 6.

| Worksheet 6 | Name: |

Let's Observe

Draw your observations of the jar and its contents.

This is what it looked like in the jar before it was covered up:	This is what it looked like in the jar after it was left for one day:

Let's Conclude

Explain the changes in the jar. What did the earthworms do?

_____
_____
_____
_____

Teacher Notes

MANAGING WASTE

LEARNING INTENTION:
Students will learn about different types of waste and how waste is managed.

SUCCESS CRITERIA:
- describe three main categories of waste
- distinguish between biodegradable waste and hazardous waste
- research and report on a method of waste disposal
- investigate how waste is managed in the local community

MATERIALS NEEDED:
- a copy of "Types of Waste" Worksheet 1 for each student
- a copy of "Biodegradable or Hazardous?" Worksheet 2, 3, and 4 for each student
- a copy of "Waste Disposal" Worksheet 5 for each student
- a copy of "Waste Management in Your Community" Worksheet 6 and 7 for each student
- pencil crayons, pencils, markers, chart paper
- clipboards (one per student)
- access to the Internet

PROCEDURE:
*This lesson can be done as one long lesson, or be divided into three shorter lessons.

1. Using Worksheets 1, 2, and 3, do a shared reading activity with the students. This will allow for reading practice and learning how to break down word parts in order to read the larger words in the text. Along with the content, discussion of certain vocabulary words would be of benefit for students to fully understand the passage.

Some interesting vocabulary words to focus on are:

- landfills
- substance
- biodegradable
- sewage
- composting
- filtration
- atmosphere
- byproduct
- hazardous
- infectious

2. Give each student a green and an orange pencil crayon, and Worksheet 4. They will categorize waste as either biodegradable or hazardous.

3. Give students Worksheet 5. They will access the internet to research a method of garbage disposal. Once students have described the method and determined the advantages and disadvantages of the method, they can share their findings with another peer or in a small group. This will allow for students to learn about other methods of garbage disposal.

4. **Arrange a class outing to the local waste management site.** Give each student a clipboard, pencil, and Worksheets 6 and 7. They will interview the site worker about waste management practices at the site. Answers to the questions can be recorded in point form or by using pictures for the students who may have difficulty with written output. *If an outing is not an available option, invite a worker from the local waste management site to come and speak to the students.

DIFFERENTIATION:
Slower learners may benefit by partnering up with a faster learner to research a method of garbage disposal on Worksheet 5. An additional accommodation is to have these learners only complete Worksheet 7 on the outing to the waste management site, omitting Worksheet 6.

For enrichment, faster learners could write a report on local waste management practices using the information they gained from the interview with the site worker.

Worksheet 1   Name:

Types of Waste

Waste can be divided into three main categories: solids, liquids, and gases. Waste must be treated differently depending on the state it is in.

Gaseous waste is the most difficult type of waste to manage. Solid and liquid waste can make gaseous waste. When solid waste rots, harmful gases are produced into the air. When liquid waste evaporates it goes into the atmosphere as gaseous waste. Once gaseous waste is in the atmosphere, there is very little that can be done to control it.

Liquid waste is more difficult to see and to manage. It is what is collected in sewers and drainage pipes, and what is sent down drains and toilets. Water is a very important substance, so it is very important that we are able to remove liquid waste from it. Filtration systems are used to do this at water treatment facilities.

Solid waste is the most visible type of waste. It is what we throw out every day, what we see as litter in the streets and in water, and what we see in landfills. Solid waste is the hardest to get rid of because it takes the longest to break down and because there is so much of it in our world.

Biodegradable or Hazardous?

Biodegradable Waste

Biodegradable waste is any type of waste that can be degraded biologically, which simply means, broken down.

Most biodegradable waste is made up of plant and animal products that can be broken down by bacteria in soil. Composting is an example.

Waste produced by the human body is biodegradable. This type of waste is the byproduct of digestion, and it is usually collected as sewage.

Inert waste is waste that cannot be broken down, but is not harmful to the environment. Some examples are sand, concrete, and chalk.

Recyclable materials are any materials that can be used again, usually after undergoing a physical or chemical change. Some examples are plastics, metals, wood, paper, and glass.

Biodegradable or Hazardous?

Hazardous Waste

Hazardous waste is any type of waste that can be harmful to living things and the environment.

Corrosive waste is a chemical that damages or destroys other living or non-living materials on contact. Batteries contain corrosive waste.

Medical waste is a type of hazardous waste. It cannot be treated as normal waste because it is usually bio-hazardous or infectious.

Radioactive waste is any waste that has radioactive chemical elements. For example, nuclear radioactive waste is extremely dangerous to living things and to the environment if not managed correctly.

Toxic waste is a chemical waste that is poisonous to living organisms.

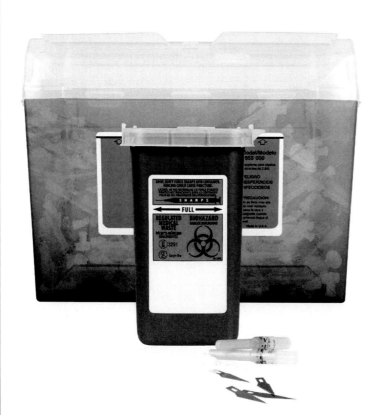

Worksheet 4   Name:

Categorize it!

Determine if the types of waste below are biodegradable or are hazardous.

Draw a **green box** around the **biodegradable** types of waste.

Draw an **orange box** around the **hazardous** types of waste.

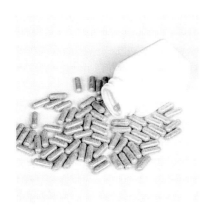  

Worksheet 5   Name:

Waste Disposal

Using the internet, research a method used to dispose of garbage. Investigate and report on:

- the method of garbage disposal
- the advantages of this method
- the disadvantages of this method

Let's Investigate!

Describe the method of garbage disposal:
___
___
___
___

Describe the advantages of this method:
___
___
___
___

Describe the disadvantages of this method:
___
___
___
___

Worksheet 6 — Name:

Waste Management in your Community

Visit your local waste management site and interview a municipal worker about how garbage is collected, sorted, and disposed for your community.

Let's Inquire!

1) How is garbage **collected**? How often is it collected?

_____
_____
_____
_____
_____
_____
_____
_____
_____
_____

2) Is the waste **sorted** at the waste management site? How?

_____
_____
_____
_____
_____
_____
_____
_____
_____
_____

| Worksheet 7 | Name: |

3) How exactly is the waste managed and *disposed* of?

_____
_____
_____
_____
_____
_____
_____

2) What are the *advantages* of this method of waste disposal?

_____
_____
_____
_____
_____
_____
_____

5) What are the *disadvantages* of this method of waste disposal?

_____
_____
_____
_____
_____
_____
_____

Teacher Notes

REDUCE, REUSE, RECYCLE!

LEARNING INTENTION:
Students will learn about ways in which materials can be reused and recycled, and methods of reduction in order to decrease waste in our world.

SUCCESS CRITERIA:
- identify objects that can be recycled or reused in order to reduce waste
- carry out plans of action to minimize waste (litterless lunch, composting, battery drive)
- make predictions, record observations, make conclusions about waste and reduction
- make a connection to the environment by identifying the benefits of waste reduction

MATERIALS NEEDED:
- a copy of "Reducing and Reusing" Worksheet 1 for each student
- a copy of "Recycling" Worksheet 2 for each student
- a copy of "Let's Go Litterless!" Worksheet 3 and 4 for each student
- a copy of "Classroom Composting" Worksheet 5, 6, and 7 for each student
- a copy of "Driving Out the Batteries!" Worksheet 8 for each student
- recyclable materials such as a pop can, aluminum foil, paper, a cardboard box, a milk carton, a plastic juice jug, a glass jar
- a large plastic bin with a lid, a garden shovel, a drill, a long stick for stirring
- vegetable and fruit scraps, egg shells, leaves, grass clippings, used coffee grinds or tea bags, or other organic material, garden soil, access to water
- pencils, pencil crayons, markers, white poster paper (optional)

PROCEDURE:
*This lesson can be done as one long lesson, or be divided into four shorter lessons.

1. Discuss with students the meaning of recycling, reusing, and reducing. Using recyclable materials, demonstrate that these are recyclable and should be sorted. Give them Worksheets 1 and 2 to complete.

2. Explain to students that they are going to plan and advertise a litterless lunch day at school, where all students are encouraged to bring a lunch with minimal packaging in order to reduce the amount of waste that is produced at school. Give students Worksheets 3 and 4. They will design a poster to advertise the event, make a prediction of the outcome, observe, and make a conclusion about waste.

3. Working as a large group, students will create and contribute to a classroom composter. Give students Worksheet 5. Read through the materials needed and what to do sections with them to ensure their understanding of the meaning of composting and the task. Gather your materials and start building! Give students Worksheets 6 and 7. They will illustrate the composting layers at the start of the project, and make observations of the decomposition of the organic matter over time on Worksheet 6. After 8 weeks, students will illustrate the compost, and respond to questions on Worksheet 7.

4. Give students Worksheet 8. They will plan and design a poster to advertise a battery drive.

DIFFERENTIATION:
Slower learners may benefit by pairing up with a peer in order to design one poster that advertises a litterless lunch event, and one poster that advertises the battery drive event.

For enrichment, faster learners could write an announcement about the litterless lunch day and the battery drive event that could be read out during the school's daily announcements.

| Worksheet 1 | Name: |

Reducing and Reusing

What exactly is reducing and reusing? Reducing is simply creating less waste to begin with, and reusing is using something for some other purpose that would have been otherwise treated as waste and disposed of.

Think **Pair** **Share**

With a partner, do some thinking and sharing of ideas of when you reduce and reuse. In the chart below, make a list of things that you reduce and reuse regularly.

Reduce	Reuse

Recycling

Recycling is an environmentally friendly way to dispose of objects that are no longer useful or needed. This helps the environment because new objects can be made from recycled materials.

Let's sort these objects into the recycling bins by drawing a line from each item to the correct recycling bin.

Worksheet 3   Name:

Let's Go Litterless!

We can also reduce our waste in the classroom in order to help the environment. Plan and advertise to have a "Litterless Lunch" day at school.

Let's Plan

Design a poster to advertise the event. Be sure to include some ideas on what would make good items to have in a litterless lunch.

What is another way you could advertise your litterless lunch idea at your school? _____

_____

| Worksheet 4 | Name: |

Let's Predict

Make a prediction about what may happen during the litterless lunch. _____

_____

Now, carry out your plan, and observe if waste in your classroom has been reduced!

Let's Observe!

Describe what happened during your litterless lunch day event.

_____
_____
_____
_____
_____
_____
_____

Let's Conclude

Was your prediction correct? Explain.

_____
_____
_____
_____
_____

Explain how reducing our waste helps the environment.

_____
_____
_____

Worksheet 5 Name:

Classroom Composting

Composting is a way to **reduce** the amount of waste that is dumped into landfill sites. By breaking down things like food waste, leaves, grass clippings, and wood bits into humus, we can put humus back into gardens, and **reuse** useful nutrients to help to grow healthy new plants.

Let's give back to the earth by creating some compost!

You'll need:

- a large plastic bin with a lid
- leaves, grass clippings
- vegetable and fruit scraps
- a long stick for stirring
- garden soil
- a garden shovel
- a drill
- access to water
- egg shells

What to do:

1. Put some soil in the bottom of the plastic bin.

2. Add some vegetable or fruit scraps, egg shells, leaves, grass clippings, and used coffee grinds or tea bags on top of the soil.

3. Moisten the soil with a little bit of water.

4. Repeat steps 1, 2, and 3 to make layers. Illustrate it on Worksheet 6.

5. Using the drill, **your teacher** will make some holes in the lid of the bin.

6. Put the lid on the bin to close it, and place it outside.

7. Use the long stick to stir your compost pile each week. Add organic materials when available. Make and record your observations of the changes you notice on Worksheet 6.

8. After several weeks, you will have created some humus that you can add to a garden to grow new plants. Illustrate it on Worksheet 7.

| Worksheet 6 | Name: |

Draw a diagram of the composter. Show the layers of ingredients you put in it.

Let's Observe!

As organic matter "cooks", explain the changes that you are noticing.

_____
_____
_____
_____
_____
_____
_____

| Worksheet 7 | Name: |

How will you use your humus?

_____
_____
_____
_____
_____
_____
_____

Illustrate the humus you have created.

Make a list of reasons why people should compost.

- _____
- _____
- _____
- _____
- _____

Challenge question:

How does composting happen in nature?

_____
_____
_____
_____
_____
_____

Worksheet 8 | Name:

Driving Out the Batteries!

Batteries contain an electrolyte which is corrosive. If corrosive and toxic matter leaches out into the environment, plants and animals can be seriously affected. This is why it is important to keep batteries out of our landfill sites and to collect and recycle them properly.

Plan and advertise to have a "Battery Drive" contest at school.

Let's Plan

Design a poster to advertise the event. Be sure to include why you are collecting batteries and how the winning class will be determined.